Prime Numbers
via
Prime Complements

Charles Harriman

Charles Harriman Publishing

Prime Numbers
via
Prime Complements

Published in the year 2014 by
Charles Harriman Publishing
Ferndale, Washington

ISBN-13: 978-0-9882408-1-0

Contents

Intro

The Prime Numbers have a new identification.

This new identification separates the Prime Numbers from their complement numbers.

These Prime Complements are easily produced.

Primes

A Prime Number is a counting number (1, 2, 3, …) with only two factors: 1 and the number itself (see Appendix A).

(1)	2	3		5	7
				11	13
				17	19
				23	
				29	31
					37
				41	43
				47	
				53	
				59	61
					67
				71	73
					79
				83	
				89	
					97

$$\vdots$$

Where are the Prime Numbers located?

Prime Locations

Except for the Prime Numbers (1), 2, and 3,
the Prime Numbers are located at one before and one after
the counting number multiples of six
(for Proof see Appendix B).

(1)	2	3	4	5	6	7
8	9	10	11	12	13
14	15	16	17	18	19
20	21	22	23	24	25
26	27	28	29	30	31
32	33	34	35	36	37
38	39	40	41	42	43
44	45	46	47	48	49
50	51	52	53	54	55
56	57	58	59	60	61
62	63	64	65	66	67
68	69	70	71	72	73
74	75	76	77	78	79
80	81	82	83	84	85
86	87	88	89	90	91
92	93	94	95	96	97

Which numbers are the Prime Complements?

Prime Complements

The Prime Complements are the **composite** numbers,
having more than two factors (see Appendix A),
that are located at one before and one after
the counting number multiples of six
(for Proof see Appendix B).

25

35

49

55

65

77

85

91

95

⋮

How are the Prime Complements produced?

Prime Complements Formation

The Prime Complements are formed by the (6a±1)(6b±1) products (with repetition of some values) for counting numbers a and b:

Prime Complements
(6a±1)(6b±1)

	5	7	11	13	17	19	23	25 ...
5	25	35	55	65	85	95	115	125 ...
7		49	77	91	119	133	161	175 ...
11			121	143	187	209	253	275 ...
13				169	221	247	299	325 ...
17					289	323	391	425 ...
19						361	437	475 ...
23							529	575 ...
25								625 ...
⋮								

Devise a new number sieve to identify the Prime Numbers!

Identification of the Prime Numbers

The Prime Numbers are identified as (1), 2, 3 and the $6n\pm1$ counting numbers which are **not** $(6a\pm1)(6b\pm1)$ products for counting numbers (1, 2, 3, …) a, b, and n:

$$\{(1), 2, 3\} \ \mathbf{U} \ \{(6n\pm1)\} \ \text{—} \ \{(6a\pm1)(6b\pm1)\} = \{\text{Prime Numbers}\}$$

("**U**" denotes sets Union, "—" denotes sets Difference)

	5	7			(1) 2 3	5	7
	11	13				11	13
	17	19				17	19
	23	25		25		23	
	29	31				29	31
(1) 2 **U** 3	35	37	—	35	=		37
	41	43				41	43
	47	49		49		47	
	53	55		55		53	
	59	61				59	61
	65	67		65			67
	71	73				71	73
	77	79		77			79
	83	85		85		83	
	89	91		91		89	
	95	97		95			97
	⋮			⋮		⋮	

A Prime Number is (1), 2, or 3, or a $6n\pm1$ counting number that is **not** a $(6a\pm1)(6b\pm1)$ product for counting numbers a, b, and n.

Appendix A

Definitions

Prime Number: A prime number is a counting number (1, 2, 3, ...) with only two factors: 1 and the number itself. Example: The prime number 17 has only two factors: 1 and 17 ($1 \cdot 17 = 17$). A prime number is one of the first three counting numbers or a $6n\pm1$ counting number which is **not** a $(6a\pm1)(6b\pm1)$ product for counting numbers a, b, and n.

Composite Number: A composite number is a counting number with more than two factors. Example: The composite number 91 has more than two factors: 1, 7, 13, 91 ($1 \cdot 91 = 91$ and $7 \cdot 13 = 91$).

The Number 1: The number 1 is the first counting number (1, 2, 3, ...) and is currently considered to be neither prime nor composite. Many current definitions of a prime number have a modified definition (soas not to affect the workings of the Fundamental Theorem of Arithmetic) that allows only two **different** factors, but the number 1 has two factors that are not different: 1 and 1 ($1 \cdot 1 = 1$). If the definition of a prime number is not modified and includes the number 1 as a prime number, then the definition of the Fundamental Theorem of Arithmetic must be modified. The Fundamental Theorem of Arithmetic states that every counting number greater than 1 is a product of prime numbers that is unique apart from the order of the prime numbers. Example: The number 90 is the product of $2 \cdot 3 \cdot 3 \cdot 5$ which is a unique list of prime numbers apart from the order ($3 \cdot 5 \cdot 2 \cdot 3$ for instance) of the prime numbers. If the number 1 is allowed to be a prime number, then the list of prime numbers for the product of prime numbers that equals the number 90 is not unique: $1 \cdot 2 \cdot 3 \cdot 3 \cdot 5$ or $1 \cdot 1 \cdot 2 \cdot 3 \cdot 3 \cdot 5$ or $1 \cdot 1 \cdot 1 \cdot 2 \cdot 3 \cdot 3 \cdot 5$ etc. (Wikipedia Contributors, "Prime Number").

Appendix B

Proofs

Theorem I. Every **prime** number ≥ 5 is of the form $6n+1$ or $6n-1$ for some integers $n \geq 1$.

Proof: Let a be any integer. By the Division Algorithm, $a = 6n+b$ for some integer n with an integer b such that $0 \leq b \leq 6$. The non-negative remainders b that are less than 6 are: 0, 1, 2, 3, 4, 5.

Hence, $a = 6n+0$ or $a = 6n+1$ or $a = 6n+2$ or $a = 6n+3$ or $a = 6n+4$ or $a = 6n+5$.

Therefore, every integer a is of the form $6n$ or $6n+1$ or $6n+2$ or $6n+3$ or $6n+4$ or $6n+5$.

The integer values for $a = 6n$, $a = 6n+2$, $a = 6n+3$, $a = 6n+4$ are **composite** for $n \geq 1$ with a having more than two factors as $6n > 2$, $6n+2 > 2$, $6n+4 > 2$, $6n+3 > 3$,

and as $6n = 2(3n)$, $6n+2 = 2(3n+1)$, $6n+4 = 2(3n+2)$, $6n+3 = 3(2n+1)$,

and as 2 divides $2(3n)$, $2(3n+1)$, $2(3n+2)$, and 3 divides $3(2n+1)$.

Thus, for $n \geq 1$, the remaining integer values for $a \geq 5$ are of the form $6n+1$ or $6n+5$ and must be the **prime** numbers ≥ 5 along with the **composite** integers > 5 of the form $6n \pm 1$

(for integers of the form $6n+5$, if $n = m-1$, then $6n+5 = 6(m-1)+5 = 6m-6+5 = 6m-1$ for some integers $m \geq 1$) ∎

Theorem II. Every product value of $(6r\pm1)(6s\pm1)$
is of the form $6n\pm1$ for some integers $n\geq1, r\geq1, s\geq1$.
Proof: The products of $(6r\pm1)(6s\pm1)$
for integers $r\geq1$ and $s\geq1$ are:
$(6r+1)(6s+1) = 36rs+6r+6s+1 = 6(6rs+r+s)+1 = 6a+1$
$(6r+1)(6s-1) = 36rs-6r+6s-1 = 6(6rs-r+s)-1 = 6b-1$
$(6r-1)(6s+1) = 36rs+6r-6s-1 = 6(6rs+r-s)-1 = 6c-1$
$(6r-1)(6s-1) = 36rs-6r-6s+1 = 6(6rs-r-s)+1 = 6d+1$
which are all of the form $6n\pm1$
for some integers $n\geq1, a\geq1, b\geq1, c\geq1, d\geq1$ ∎

Theorem III. Every product value of $(6r\pm1)(6s\pm1)$
for integers $r\geq1$ and $s\geq1$ is **composite**.
Proof: Let $(6r\pm1)(6s\pm1) = x$ for some integer x
and for some integers $r\geq1$ and $s\geq1$.
Then x has more than two distinct factors:
$1, x, (6r\pm1)$ for $r\geq1$, and $(6s\pm1)$ for $s\geq1$.
Therefore, by the definition of composite number,
every product value of $(6r\pm1)(6s\pm1)$
for integers $r\geq1$ and $s\geq1$ is **composite** ∎

Definition of even number: For any integer x, x is defined as
an **even** number if and only if $x = 2k$ for some integer k
(x divided by 2 equals some integer and no remainder).

Definition of odd number: For any integer x, x is defined as
an **odd** number if and only if $x = 2m+1$ for some integer m
or if and only if $x = 2n-1$ for some integer n
(x equals some even integer plus 1, or x equals some even
integer minus 1).

Theorem IV. For the products of two integers, even•even, even•odd, odd•even, and odd•odd, only odd•odd produces an odd number.

Proof: Let a and b be any two integers. Then the product of $a•b$ is either even•even or even•odd or odd•even or odd•odd as an integer is either even or odd.

For a an even number and b an even number $a = 2(h)$ and $b = 2(w)$ by the definition of an even number for some integers h and w, and for a an odd number and b an odd number $a = 2m+1$ and $b = 2t+1$ by the definition of an odd number for some integers m and t.

Then, $a•b = $ even•even $= 2(h)•2(w) = 2(2h•w) = 2(c)$ which is even by the definition of an even number for some integer c, or $a•b = $ even•odd $= 2(h)•(2t+1) = 4h•t+2h = 2(2h•t+h) = 2(k)$ which is even by the definition of an even number for some integer k, or $a•b = $ odd•even $= b•a = $ even•odd by the commutative property of multiplication, which is even, or $a•b$ $= $ odd•odd $= (2m+1)•(2t+1) = 4m•t+2m+2t+1 = 2(2m•t+m+t) +1$ $= 2g+1$ which is **odd** by the definition of an odd number for some integer g.

Thus, for the product of two integers $a•b$ (even•even, even•odd, odd•even, and odd•odd) only odd•odd produces an **odd** number for some integers a and b ∎

Theorem V. Every composite number for the integers of the form $6n\pm1$ for integers $n\geq1$ is a product value of $(6r\pm1)(6s\pm1)$ for integers $r\geq1$ and $s\geq1$.

Proof: Assume there exists an integer x such that x is composite and of the form $6n\pm1$ for integers $n\geq1$ such that x is **not** a product value of $(6r\pm1)(6s\pm1)$ for integers $r\geq1$ and $s\geq1$. Then x is odd as $x = 6n\pm1 = 2(3n)\pm1 = 2k\pm1$ which is odd by the definition of an odd number for some integers $n\geq1$ and $k\geq3$. Therefore, $x = 2g+1$ or $x = 2h-1$ which are both odd by the definition of an odd number for some integers $g\geq3$ and $h\geq3$. Since x is composite, then x has more than two factors by the definition of a composite number to have $x = 1 \cdot x$ and $x = a \cdot b$ for some integers a and b with $a\neq1$ and $b\neq1$.

For the products of two integers, even·even or even·odd or odd·even or odd·odd, only odd·odd produces an odd number which is proved by Theorem IV.

Hence, with x an odd number, $x = 6n\pm1$ for some integers $n\geq1$, and $x = a \cdot b$ for some integers $a\geq5$ and $b\geq5$, then a is odd and b is odd with $a = 2c\pm1$ and $b = 2d\pm1$ by the definition of an odd number for some odd integers $a\geq5$ and $b\geq5$ and for some integers $c\geq3$ and $d\geq3$.

Thus, $x = 6n\pm1 = a \cdot b =$ odd·odd for some integers $n\geq1$ and for some odd integers $a\geq5$ and $b\geq5$
$= (2c+1)(2d+1)$ or $(2c+1)(2d-1)$ or $(2c-1)(2d+1)$ or $(2c-1)(2d-1)$ for some integers $c\geq3$ and $d\geq3$.

But, every $6n\pm1$ composite number of $a \cdot b$, which is odd·odd for some integers $n\geq1$ and for some odd integers $a\geq5$ and $b\geq5$, is a product value of $(6r\pm1)(6s\pm1)$ as
$(2c+1)(2d+1) = 4cd+2c+2d+1 = 36rs+6r+6s+1 = (6r+1)(6s+1)$,
$(2c+1)(2d-1) = 4cd-2c+2d-1 = 36rs-6r+6s-1 = (6r+1)(6s-1)$,
$(2c-1)(2d+1) = 4cd+2c-2d-1 = 36rs+6r-6s-1 = (6r-1)(6s+1)$,
$(2c-1)(2d-1) = 4cd-2c-2d+1 = 36rs-6r-6s+1 = (6r-1)(6s-1)$,
for $c = 3r$ and $d = 3s$ for some integers $r\geq1$ and $s\geq1$.
This is a contradiction of the original assumption.
Therefore, the assumption is false and the statement of Theorem V is true ∎

List of Primes
(1), 2, to 599

(1),

2, 3, 5, 7, 11, 13, 17, 19, 23, 29, 31,

37, 41, 43, 47, 53, 59, 61, 67, 71, 73,

79, 83, 89, 97, 101, 103, 107, 109, 113,

127, 131, 137, 139, 149, 151, 157, 163,

167, 173, 179, 181, 191, 193, 197, 199,

211, 223, 227, 229, 233, 239, 241, 251,

257, 263, 269, 271, 277, 281, 283, 293,

307, 311, 313, 317, 331, 337, 347, 349,

353, 359, 367, 373, 379, 383, 389, 397,

401, 409, 419, 421, 431, 433, 439, 443,

449, 457, 461, 463, 467, 479, 487, 491,

499, 503, 509, 521, 523, 541, 547, 557,

563, 569, 571, 577, 587, 593, 599

Primes & Prime Complements
(1), 2, to 127

(1)	2	3	5	7
			11	13
			17	19
			23	25
			29	31
			35	37
			41	43
			47	49
			53	55
			59	61
			65	67
			71	73
			77	79
			83	85
			89	91
			95	97
			101	103
			107	109
			113	115
			119	121
			125	127

Primes
(1), 2, to 127

(1)	2	3	5	7
			11	13
			17	19
			23	
			29	31
				37
			41	43
			47	
			53	
			59	61
				67
			71	73
				79
			83	
			89	
				97
			101	103
			107	109
			113	
				127

Prime Complements
(5...25)(5...25)

	5	7	11	13	17	19	23	25
5	25	35	55	65	85	95	115	125
7		49	77	91	119	133	161	175
11			121	143	187	209	253	275
13				169	221	247	299	325
17					289	323	391	425
19						361	437	475
23							529	575
25								625

Primes & Prime Complements
131 to 247

Primes
131 to 241

131	133	**131**	
137	**139**	**137**	**139**
143	145		
149	**151**	**149**	**151**
155	**157**		**157**
161	**163**		**163**
167	169	**167**	
173	175	**173**	
179	**181**	**179**	**181**
185	187		
191	**193**	**191**	**193**
197	**199**	**197**	**199**
203	205		
209	**211**		**211**
215	217		
221	**223**		**223**
227	**229**	**227**	**229**
233	235	**233**	
239	**241**	**239**	**241**
245	247		

Prime Complements

(29...49)(5...19)

	29	31	35	37	41	43	47	49
5	145	155	175	185	205	215	235	245
7	203	217	245	259	287	301	329	343
11	319	341	385	407	451	473	517	539
13	377	403	455	481	533	559	611	637
17	493	527	595	629	697	731	799	833
19	551	589	665	703	779	817	893	931

Primes & Prime Complements
251 to 367

251	253
257	259
263	265
269	**271**
275	**277**
281	**283**
287	289
293	295
299	301
305	**307**
311	**313**
317	319
323	325
329	**331**
335	**337**
341	343
347	**349**
353	355
359	361
365	**367**

Primes
251 to 367

251	
257	
263	
269	**271**
	277
281	**283**
293	
	307
311	**313**
317	
	331
	337
347	**349**
353	
359	
	367

Prime Complements

(53...73)(5...13)

	53	55	59	61	65	67	71	73
5	265	275	295	305	325	335	355	365
7	371	385	413	427	455	469	497	511
11	583	605	649	671	715	737	781	803
13	689	715	767	793	845	871	923	949

Primes & Prime Complements
371 to 481

Primes
373 to 479

371	373		373
377	379		379
383	385	383	
389	391	389	
395	397		397
401	403	401	
407	409		409
413	415		
419	421	419	421
425	427		
431	433	431	433
437	439		439
443	445	443	
449	451	449	
455	457		457
461	463	461	463
467	469	467	
473	475		
479	481	479	

Prime Complements
(77...97)(5...7)

	77	79	83	85	89	91	95	97
5	385	395	415	425	445	455	475	485
7	539	553	581	595	623	637	665	679

Primes & Prime Complements
485 to 595

Primes
487 to 593

485	487		487
491	493	491	
497	499		499
503	505	503	
509	511	509	
515	517		
521	523	521	523
527	529		
533	535		
539	541		541
545	547		547
551	553		
557	559	557	
563	565	563	
569	571	569	571
575	577		577
581	583		
587	589	587	
593	595	593	

Prime Complements

(101…121)(5)

101	103	107	109	113	115	119	121

5	505	515	535	545	565	575	595	605

Resources

Burton, David M. *Elementary Number Theory*.
    Wm. C. Brown Publishers, Dubuque, IA, 1994.

Epp, Susanna S. *Discrete Mathematics With Applications*.
    2nd edition. PWS Publishing Co., Boston, MA 1995.

Harriman, Charles. *Prime By Secret Design*. 2nd edition.
    Harriman Publishing, Bellingham, WA, 2008.

Wells, David G. *Prime Numbers: The Most Mysterious Figures
    In Math*. John Wiley & Sons, Inc., Hoboken, NJ, 2005.

Wikipedia Contributors. "Prime Number." *Wikipedia:
    The Free Encyclopedia*. Wikimedia Foundation,
    3 September 2014. Web. 9 September 2014.
    www.wikipedia.org